SCIENCE WITH A BEAT

I0469740

THE A-B-C SEA CRITTER BOOK

WRITTEN BY JACQUIE HAWKINS

PART OF THE A-B-C SCIENCE SERIES

A

10 inches tall

An **ANEMONE**
looks just
like a flower
although it's not a plant.
Instead it is an animal,
although a little weird
I grant.
It has a tube-like body
and its mouth is
at the top,
Surrounded by a crown of tentacles that sting a lot.
You'll find one in the ocean, both Atlantic and Pacific.
It grows up to ten inches and it really looks terrific.

The **ARCTIC TERN'S** a
graceful bird, that's
mostly gray and white.
Its wings are very
slender and to
watch one is a sight.
It glides through air then
swoops and dives in
water with a swish
And when it comes
back out again…
behold, it has a fish!
It lives a very long

time……even up to thirty years.
It loves to live where it is cold,
and that is very clear.

2.

ACORN BARNACLES

stick themselves to
hard things just like glue;
To wharfs, ships, rocks
and even
animal shells....
without a thank-you.
They have cone-shaped
bodies with a
trap door on the top.
When covered up
with water that
trap door will
open up.

Plankton's what they gobble up...
For them that's pretty smart.
Their heads are down...their feet are up.
They have one eye....and have no heart!

16 -26 inches
long

The
**ATLANTIC
MACKERAL**
(two feet long)
Moves about
in quite large
schools.
It travels north
in the spring
to breed
Then goes
back
south through the winter's freeze.

53 inches long
wingspan up to 11 feet

The **ALBATROSS**,
a wanderer,
spends most
of its life exploring,
Flying as much as
three hundred miles
in one day
while just soaring.
It only comes to
land to lay its
eggs and feed
its young.
It feeds on fish and
squid plus food
that from ships have been flung.

B

BAT STARS are
a kind of starfish.
They come in many
different colors.
Some are solid,
some are mottled,
some are bright and
some much duller.
Most of them
have five arms but
they can have up to nine.
Light sensors that detect
their prey at each end's
what you find.

The stomach juices ooze out which will liquefy their prey.
Then they simply slurp it up without little delay.

4.
BULLET-HEAD
PARROT FISH
start life
as females
that are striped;
Three stripes that
are very black
and three stripes
that are white.
Later some turn
into males

but only just a few.
Yet all become quite colorful with
sharp teeth that sure can chew!
While eating algae they crunch coral…
which I think is weird.
One fish can turn five hundred pounds of
coral to sand in just one year!

up to 16 feet long

BELUGA WHALES
or White Whales,
(often called the
Sea Canaries)
squeak
and squeal
and click
and whistle.

They're quite legendary.
They are gray or brown at birth but turn white at age five.
They stick together in huge pods.
That's how they can survive

The brilliant
pattern
on male
BOXFISH
really is
impressive.
It's like an
artist's
painting arm became a bit excessive.
It can release a deadly toxin if it's under stress.
This can kill somebody so be careful, I suggest!

The **BLUE RING
OCTOPUS** is quite
dangerous, yes indeed.
You'd find one
near Australia on the
Great Barrier Reef.
It is pretty small…just
about a golf ball's size.
It has a painless bite
and yet to get stung
is not wise.
Its venom is so
poisonous that
death's a sure result.
One bite's enough to kill
up to twenty six adults!!
It's normally pretty
shy, will change its color

or will hide.
But, if its blue rings light up…
then to flee you should decide!

6.

The **BRITTLE STAR**
is not a fish.
It has no gills or fins.
It has five arms like
starfish do
but it has a
bumpy skin.
It glides along the
rocks and sand
on tiny tube-like feet.
Of course, it eats food
with its mouth....but its

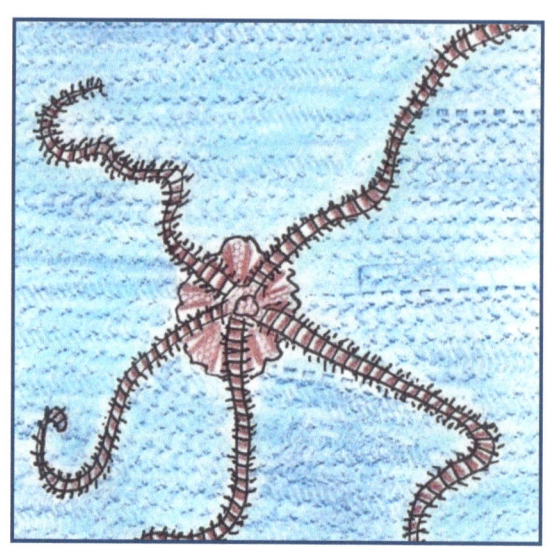

mouth is underneath!

BASKET STARS have
arms but they branch
out in feathery shapes.
When they are scared they
lift them up;
around themselves
they're draped.
They look much like a
basket then.
That's how they
got that name.
They have five jaws
around their mouth!
That simply is insane!
They can live to
thirty-five years and weigh to eleven pounds.
They live in mud at the bottom;
can be yellow, white or brown.

up 100 feet long

The <u>largest</u> <u>animal</u>
<u>on</u> the <u>earth</u>
<u>is</u> the <u>great</u> **BLUE**
WHALE.
<u>It</u> can <u>weigh</u>
two-<u>hundred</u> <u>tons,</u>
be a-<u>hundred</u> feet
<u>long</u> from
<u>head</u> to <u>tail</u>.

Its <u>tongue</u> weighs as <u>much</u> as an <u>elephant</u>!
It's <u>heart</u>, as <u>much</u> as a <u>car</u>!
Its <u>baby</u> can <u>weigh</u> three <u>tons</u> at <u>birth</u>
and <u>that's</u> just <u>plain</u> bi<u>zarre</u>!
<u>It</u> can <u>eat</u> four <u>tons</u> of <u>Krill,</u> and <u>that's</u> in <u>just</u> one <u>day</u>!
Its <u>loud</u> groans, <u>pulses</u> <u>and</u> weird <u>moans</u>
can be <u>heard</u> a <u>thous</u>and <u>miles</u> <u>away</u>!

C

SEA <u>CUCUMBERS</u> are
<u>long</u> and <u>round</u>ed,
<u>kind</u> of <u>like</u> a cu<u>cumb</u>er.
<u>But</u> they <u>do</u> not
<u>taste</u> like <u>one,</u>
<u>just</u> in <u>case</u> you
<u>wondered</u>.
<u>When</u> they <u>eat</u>
they <u>take</u> food <u>in</u> and
<u>through</u> a <u>tube</u>
it's <u>sent</u>
Un<u>til</u> it <u>finally</u>

10 inches long

<u>makes</u> its <u>way</u> <u>to</u> the <u>other</u> <u>end</u>.

A <u>strange</u> <u>look</u>ing <u>creature</u> <u>is</u> one <u>called</u>
CLOWN <u>TRIGGER</u>FISH.
<u>I</u> don't <u>think</u> I <u>ever</u> <u>saw</u> a <u>fish</u> that <u>looked</u> like <u>this</u>!

<u>To</u> protect
their <u>bodies</u>,
<u>most</u> **CRABS**
<u>have</u> a
sturdy <u>shell</u>
<u>And</u> a <u>set</u> of
<u>pincers</u> <u>on</u>
their <u>front</u> two
<u>legs</u> as <u>well</u>.
But <u>Hermit</u>
<u>Crabs</u> do <u>not</u>
have <u>any</u>

<u>hard</u> shells <u>of</u> their <u>own</u>.
They <u>find</u> an <u>empty</u> <u>shell</u>, move <u>in</u>, and
<u>call</u> that <u>shell</u> their <u>own</u>.

The **CORMORANT**,
to catch its prey,
into the water dives.
And when it's
through it
spreads its wings
straight out before
it flies.
It's like it has a
clothesline where its
wings hang out to dry.
Once its wings are
dry enough the
bird takes to the sky.

The **COPPERBAND
BUTTERFLY
FISH** actually
has a beak
Or snout that
reaches
crevices so that
critters it can
reach.
It has an
eyespot
near its tail to
confuse its
predators.

7 inches long

You'll find one near a reef or all
along a rocky shore.

The **CHRISTMAS TREE WORM**
kind of looks
just like a pretty Christmas tree.
It has two crowns
of spiral shapes.

It's found
out on a
coral reef.

Most of its body
stays inside the
coral where it
likes to hide.
The 'tree' part is
what divers see
for that's what's on
the outside.

Some are orange,
yellow, red or
blue and some
plain white,
But though they are
very small,
to see one is a
pretty sight!

D

7 to 8½ feet

DOLPHINS
use the sonar
waves
by making
clicking
sounds.
They listen for
the echo and
know if fish
are around.

But when they talk to one another they
speak by making squeals.
I wouldn't understand them but for
them that is ideal.

The strangely
colored
DOLPHINFISH
is easy enough
to track
Because of its
large, dorsal fin
that runs along
its back.
It lives in

6 feet long

tropical water, eats crustaceans, squid and fish.
It usually travels in small schools.
If not, I bet that'd be their wish.

12.

A **DOVEKIE** is a
bird that
loves the cold
and so lives way
up north.
It's black and
white, is
stubby looking,

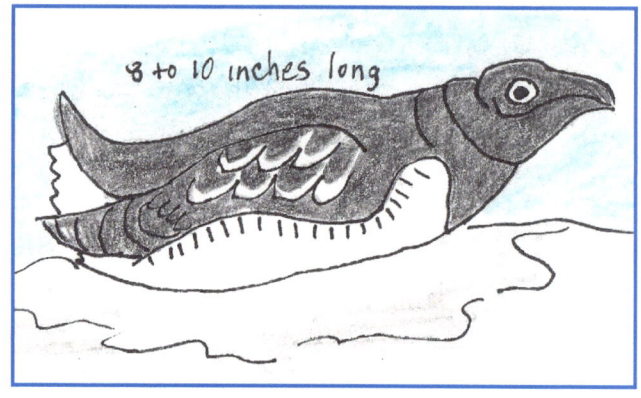

8 to 10 inches long

and has a tail that's short.
It lives up in the Artic and it breeds along the coast.
It moves a little south for winter.
Because to them that's warm as toast.

E

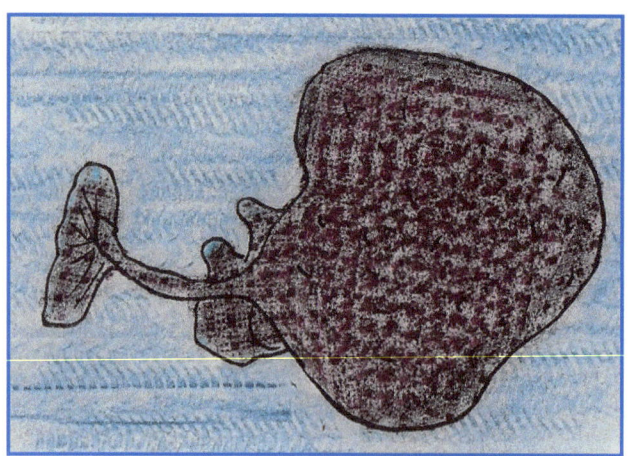

Large **ELECTRIC
RAYS** produce,
at times two-
hundred volts,
Enough to stun
a diver.
That surely is
no joke.
They're also
called
Numbfish and

Crampfish.
Bet you can guess why!
Luckily it's a bottom dweller and most of them are shy.

The **ELLIPTICAL STAR**
Or the Star Coral
grows up to
one foot wide.
It's a colony of
soft bodied animals.
The coral is on
the outside.
It has a partnership
with algae, so they
help each other out.
The algae provides

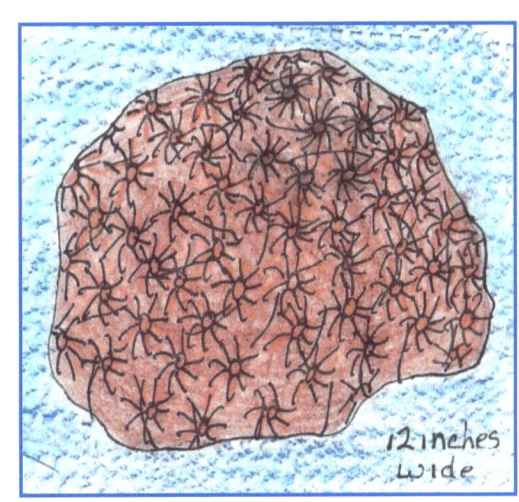

12 inches wide

the food it eats
and the star provides the house.

45 inches tall

The largest penguin,
the **EMPEROR PENGUIN,**
is mostly black and white.
It spends its life in a
humongous colony
on the Antarctic ice.
The mama lays her
egg in winter and
then heads out to sea.
For sixty days the
egg's kept warm
on top of papa's feet.
Mom returns when the
egg is hatched so
papa can then eat.
He is very hungry
and is very glad to leave!

F

22 feet long

To escape his enemies the
FLYING FISH
takes off.
Up to forty miles per hour
this fish
can blast off.

Using his wing-like fins this fish can glide
three-hundred feet,
and up to five feet in the air!
That is quite a feat!

The **FEATHER FAN WORM'S**
feathery tentacles
look much like
a fan.
They trap small bits
of seaweed
and animals.
At least that
is its plan.

Deep down in
the ocean where it
is as black as ink,
The
**FLASHLIGHT
FISH** lights up the
dark.
Its eyes
go blink…blink…
blink.

It has little color and is just twelve inches long.
So all you see is blink…blink…blink
as it swims along.

2 feet long

The **FOOTBALL
FISH**
looks kind of like
a football
in the sea.
It lives way down
where it is dark
where no prey
it can see.
And so it has a
'fishing rod' with
a light right
at the end.

It attracts its prey as through the darkness it descends.

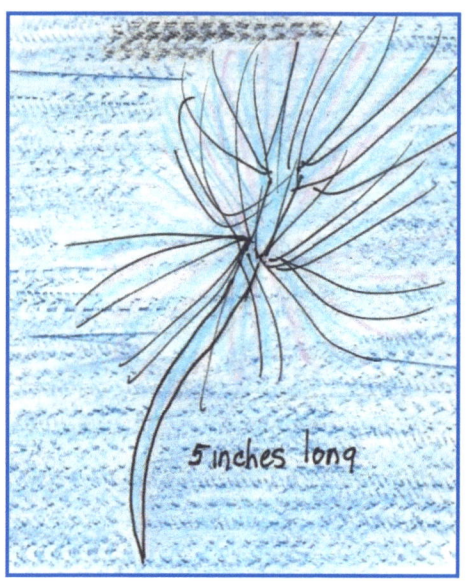

5 inches long

The **FEATHER DUSTER WORM** really does not dust at all. It is just five inches long and that is rather small. It hides inside its flexible tube that's made up of glued sand. It catches food with its feathery gills and eats all that it can.

G

The **GREEN SEA TURTLE** can travel up to almost three-thousand miles. It lays its eggs in its nesting grounds. That must take quite a while.

One can weigh seven-hundred pounds. It has green-colored skin and uses its paddle-like flippers to dig in the sand to help it swim.

The <u>largest</u> <u>clam</u>
that's <u>in</u> the <u>world</u>
<u>is</u> the **GIANT CLAM**.
<u>It</u> makes <u>food</u> by
<u>photosyn</u>thesis
<u>just</u> like <u>green</u>
plants <u>can</u>.
It <u>also</u> <u>filters</u>
<u>plank</u>ton <u>from</u> the
<u>water</u> <u>like</u> a <u>sieve</u>.
<u>That</u> way <u>it</u> gets
<u>lots</u> to <u>eat,</u> which
<u>helps</u> the <u>clam</u> to <u>live</u>.

The **GANNET** <u>is</u> an
<u>expert</u> <u>diver</u>,
<u>I</u> have <u>little</u> <u>doubt</u>.
It <u>has</u> a <u>flap</u> in
<u>front</u> of its <u>nose</u> to
<u>keep</u> sea <u>water</u> <u>out</u>.
It's <u>mostly</u> <u>white</u>
with a <u>yellow</u>ish <u>head</u>
and <u>black</u> tips
<u>on</u> its <u>wings</u>.
<u>It</u> can <u>dive</u> to
<u>ninety</u>-eight <u>feet</u>
and can
<u>catch</u> 'most
<u>anyth</u>ing!

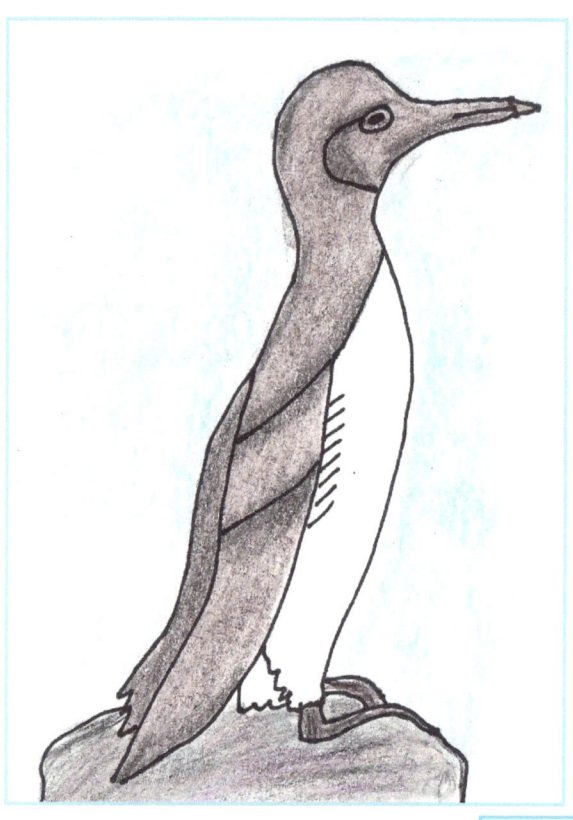

The **GUILLEMOT**,
like most seabirds,
has two duck-like ,
webbed feet.
But unlike some
it rarely ever
dives to get some
fish to eat.
It is known,
at times, to steal
the food from
other birds.
That's not nice for
a bird to do…
of that please
be assured.

The **GOOSE BARNACLES**
does not look much
like a goose….
well, not to me.
Instead it looks
more like the lightbulbs
found around a
Christmas tree.
It floats around
on things like logs
and boats out in the sea.
Six pairs of feathery
arms poke out to
collect the food it eats.

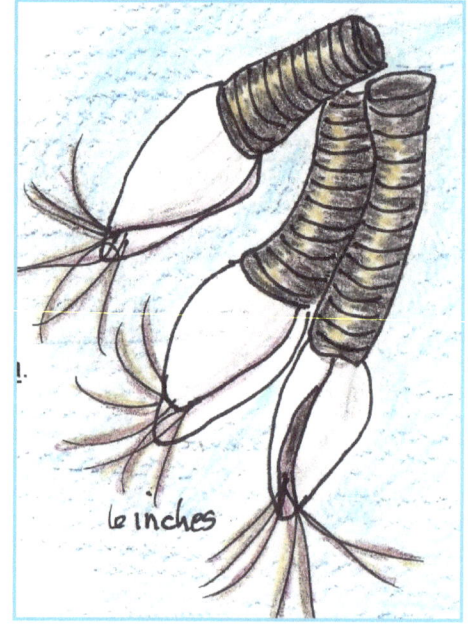

6 inches

The **GOOSEFISH,**
sometimes called
the Angler,
looks quite
strange to me.
It stays hidden
If it can
at the bottom
of the sea.
But a flap of
skin sticks out that
fish think must
taste good.

1-6 feet long

When fish come close he snaps them up.
That's how he catches food.

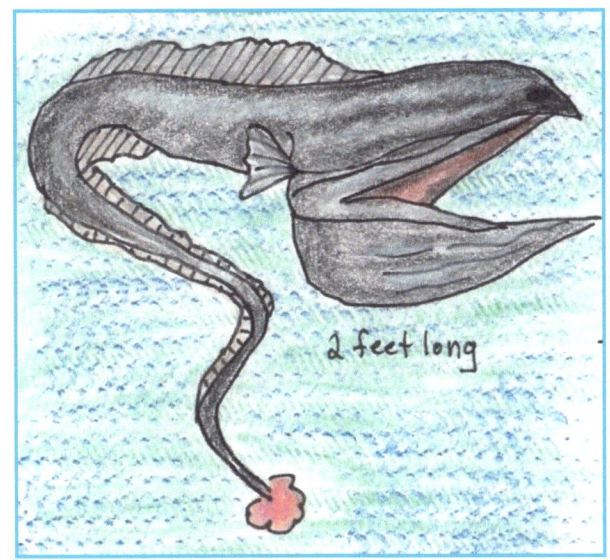

2 feet long

The **GULPER EEL**
does not look like
the other eels
around.
Its mouth is
way too big for him
and needs to be
scaled down.
It can swallow
fish whole, even
fish two times the
size he is.
That head of his is
so big!

It can't possibly be his!

H

HUMPBACK WHALES

spend all their
summers feeding at
one of the Poles
Because the food is
rich there and to
eat krill is their
only goal.
In winter they go to
the tropics
where it's very warm
So they can breed
and very soon

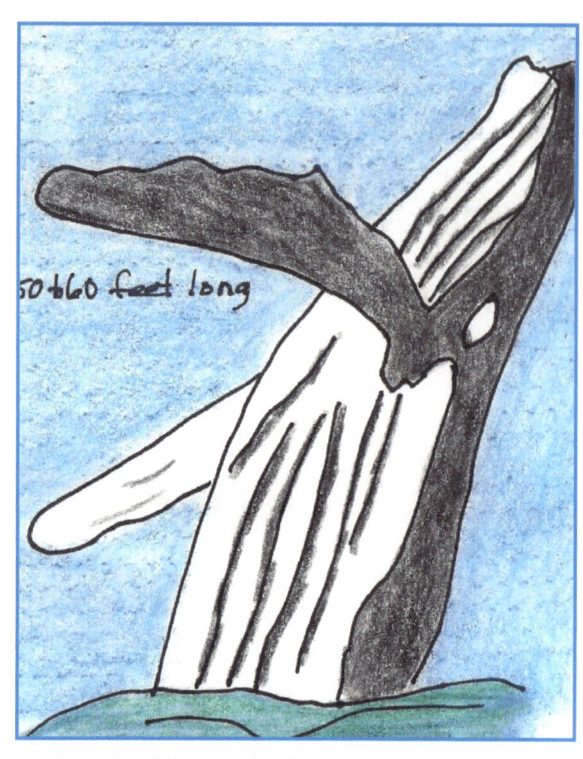

a little Humpback Whale is born.
Mom (just like with humans) feeds her baby her own milk
Until the babe is old enough and can catch its own krill.

The **HAMMERHEAD SHARK'S** head
is quite flat and
also very wide.
The front part's long and
broad and it has
one eye on each side.
It can't see things
right in front and
so might just collide
Unless, as he is
swimming,
He swings his head
from side to side.

The **HARLEQUIN TUSKFISH**
is a fish that's about
twelve inches long.
Its eyes move
independently
and that seems
very wrong.
It's found in the
Pacific west,
mainly on coral reefs.
Its eyes are
dark red-orange,
and, I swear…….
it has BLUE teeth!

The **HATPIN URCHIN**
is the largest
urchin
ever known.
It is up to a foot
long…
when it is fully-
grown.
Its 'hat pins' are
long and sharp.
If you're stuck
you will cry.
Its spines

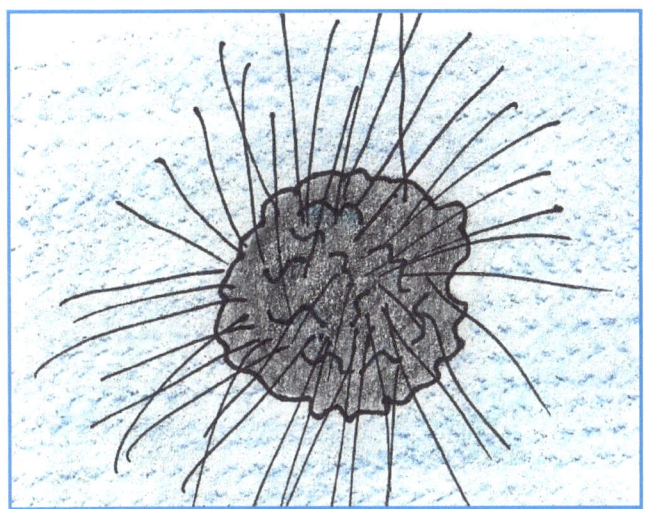

release a venom…so you could possibly die.

The
HATCHETFISH
is pretty small…
three inches,
maybe less.
All day it stays
in the deep,
dark water,
I guess so
it can rest.
But at night
up to the surface it
swims so it can eat.
A row of lights along its belly
gives off a bluish streak.
This confuses his predators in terms of its body size.
They will think he's so much bigger and the
truth won't recognize.

The **HAWKSBILL
TURTLE** has a
hawk-like jaw
and a
beautiful shell.
Sometimes it will eat
poisonous sponges
but you can
never tell.
For they do not
seem to hurt
or faze him
in the least.
He usually stays

under the water and comes up just to breathe.

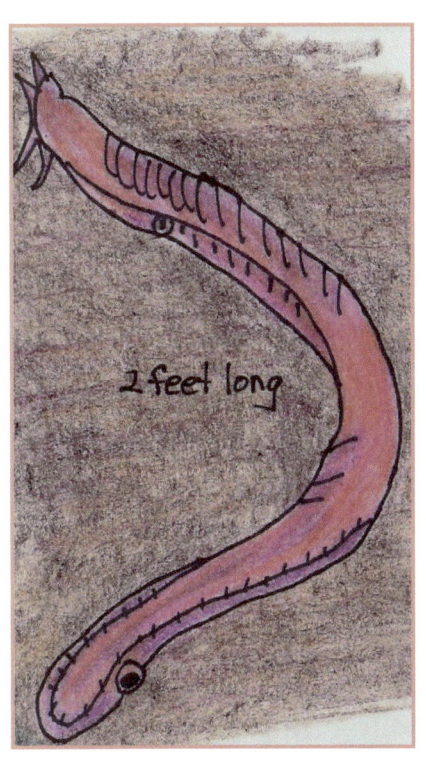

2 feet long

The **HAGFISH** doesn't
look like a fish…
not one I ever saw.
It has no scales,
looks like a snake and
doesn't even have a jaw!
Its mouth is just a slit
with lots of tentacles
all around.
Buried in soft seabed mud
is where it can be found.

The **HORN SHARK** is a small shark,
only three feet long at most.
It's found in the Pacific Ocean along California's coast.
It is a bottom dweller and it doesn't move too fast.
It is pretty harmless…unless it is harassed.

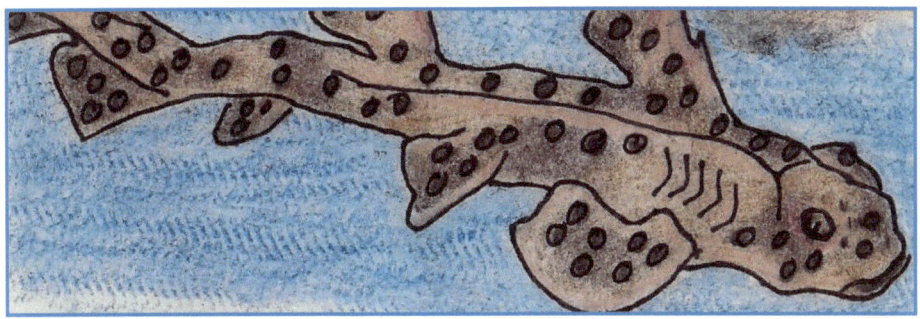

I

The marine **IGUANA'S** the only lizard that
spends its life at sea
Swimming close to islands in its search to eat seaweed.
When it dives its heart rate slows, less oxygen it needs.
Its tail is very powerful and he swims at a great speed.

The Giant **ISOPOD'S** related to the
very small wood louse.
It looks a lot like one but it's about the size of a mouse!

ICEFISH live
where it
is cold
along the
Antarctic
coast.
They live
where there's
ice all year
long.

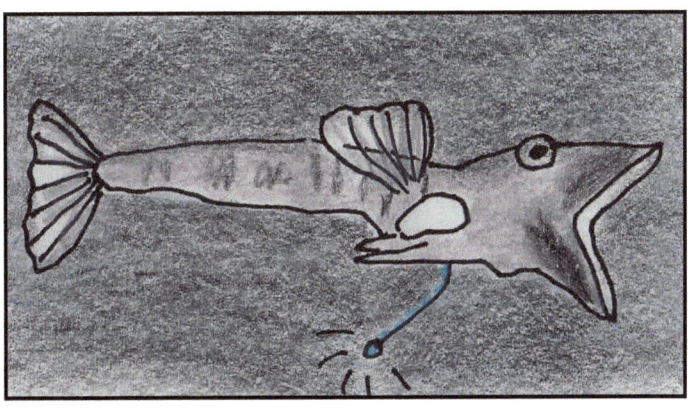

Further south they'd simply roast.
Because they do not have red blood cells,
colorless they appear.
Why you can see right through them…
like ice, their body's clear!

up to
18 inches
long

The **IVORY GULL**
is a bird
that totally is white…
Though its legs
and eyes are black.
It's pretty
when in flight.
It's a scavenger and
so it eats what
dead things
it can find.
It's known to follow
Polar Bears and
eat what's left behind.

26.

J

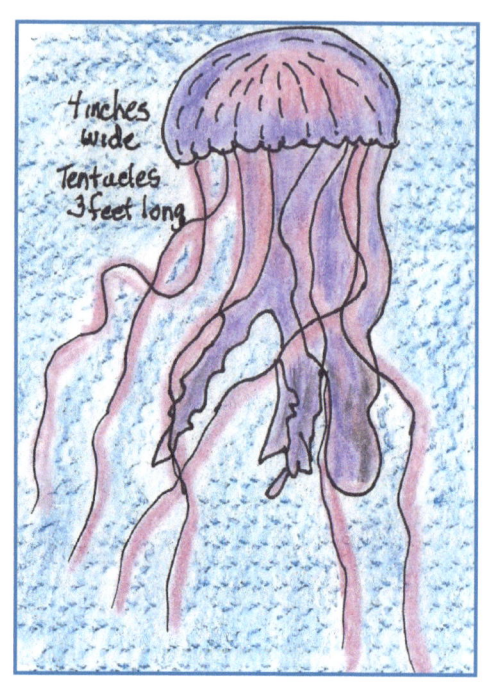

JELLYFISH are
soft-bodied creatures that
you can see
right through.
They'd feel a lot like jelly
but I wouldn't advise you to.
They drift along on the
current with their
tentacles
dangling down.
Please be aware!
These tentacles are where the stinging cells are found.
If you're ever stung by one I doubt you will forget
And the next time you go swimming you'd be careful.
That's my guess.

K

The **KILLER WHALE**
has pointed teeth that
are quite sharp,
I'm told.
But if its prey is
small enough,
why, he'll just swallow
that prey whole.
A Killer Whale eats
seabirds, fish and
seals from head to tail
But it's known at times to even eat another whale!

L

LINED <u>SWEET</u>LIP
F<u>ISH'S</u> <u>lips</u> are <u>flashy</u>,
<u>fleshy</u> <u>and</u> bright <u>yellow</u>.
It's <u>like</u> they <u>put</u> on
<u>yellow lip</u>stick,
<u>whethe</u>r a <u>gal</u> or <u>fellow</u>.
The <u>youngs</u>ter <u>fish</u> do
<u>not</u> look <u>like</u> their <u>mothers</u>
or their <u>dads</u>.

They <u>have</u> a <u>different col</u>or
<u>and</u> with <u>spots</u> or <u>bands</u> are <u>clad</u>.
A<u>dult</u> fish <u>will</u> get <u>more</u> and <u>more</u> stripes
<u>as</u> they <u>grow</u> and <u>age</u>.
<u>Hav</u>ing <u>yellow lips</u> and <u>lots</u> of <u>stripes</u> sure <u>seems</u> the <u>rage</u>.

A <u>seahorse called</u> the
<u>LEAFY</u> SEA <u>DRAGON</u>
is <u>very strange</u> indeed.
It <u>lives</u> and <u>hides</u> a<u>mong</u>
the <u>leafy</u> <u>patch</u>es
o<u>f</u> sea<u>weed</u>.
It <u>hides</u> from
hun<u>gry predators</u>,
<u>and</u> to <u>find</u> him...
<u>they</u> will <u>try</u>,
But he <u>looks</u> just <u>like</u> a
<u>piece</u> of <u>seaweed</u>
<u>that</u> is <u>float</u>ing <u>by</u>.

28.

LOBSTERS hunt at night and hide among the rocks all day. They have two really big pincers used to crack apart their prey. They can live up to 70 years…

and don't slow down with age.
The largest lobster recorded…
over forty pounds it weighed!
They do not swim but walk along the bottom of the sea
But they can swim backwards if they're trying hard to flee.

LIMPETS live on algae. On a lot they like to dine. Just like snails they leave a sticky mucus trail behind.

They use the trail so they can always follow it back home.
Like snails their homes protect them
For their shells are hard like stone.

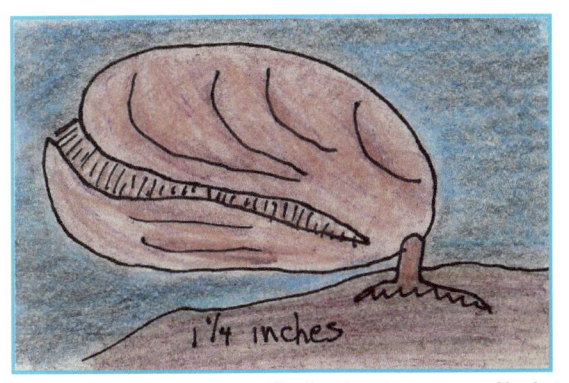

1 1/4 inches

I don't know why
the tiny **LAMPSHELL**
ever got that name
Because it and
a lamp do not look
in the least the same.
Unlike some
deep water fish that carry lights, this one does not
But if it did (just one inch long) it would not shine a lot.
Its kind has lived six hundred million years on earth...
It's true!
It has two shells that open.
Hairy tentacles gather food.

15 inches
long

The
LIONFISH
is one of
the most
unusual fish
found on
a reef.
The spines
found on its
fanlike fins
protect it.
Not much
gives it grief.
Its back
spines are quite poisonous and are used for defense...
but rarely used to kill its prey.
To me that makes no sense.

11 feet long

The **LEOPARD SEAL**
is slender and
is truly built
for speed.
It's powerful,
aggressive and
a very
curious breed.
It's a hunter,
second only
to the
Killer Whale.
It's said they've stalked, attacking humans,
but that might be a tale.

M

The slow-moving
MANATEE,
sometimes called
Sea Cow,
Stays in shallow,
coastal waters as
through sea-grass it plows.
That's like an underwater
forest or meadow
in the sea.
It eats thirty pounds a day!
That seems a lot to me!
Though it's called a
Sea Cow,
I doubt that it can moo
And I doubt it lays around
all day its cud to chew.

up to
10 feet
long

When the tide goes out to sea,
MUDSKIPPERS burrow in
And stay beneath the sand until the tide comes in again.
These fish can breathe no matter if they're in water or out.
They 'skip' across the mud using their fins to move about.

To think that
MUSSELS
have big muscles
really would
be wrong.
Yet their two shells
are hard to open.
So I guess folks
say they're strong.
They open up their
shells a little so that food can get in,
But once it's in, just like a trap,
they snap back shut again.
To protect themselves from waves
as the tide goes in and out,
They stick hard to a rock so that they
do not bounce about.

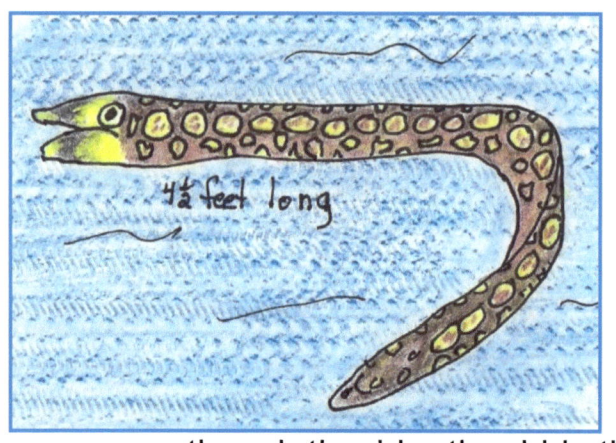

4½ feet long

There <u>are</u> a
<u>hundred</u>
<u>different</u> <u>species</u>
<u>of</u> the
MORAY EEL.
It's <u>shy</u> when <u>near</u>
to <u>humans</u>, and
<u>that</u> is <u>a</u> big <u>deal</u>.
Its <u>reputation</u>
<u>is</u> not <u>good</u>,
though they'd <u>rather</u> <u>hide</u> than <u>fight</u>.
But you'll <u>have</u> to <u>pry</u> them <u>off</u> of <u>you</u>…
if <u>you</u> they <u>ever</u> <u>bite</u>.

The **MANTA RAY**
looks a <u>lot</u>
like <u>batman</u>.
<u>That's</u> if <u>you</u>
ask <u>me</u>.
But <u>it's</u> a
<u>whole</u> lot <u>bigger</u>.
The <u>largest</u> –
<u>twenty</u>-two <u>feet</u>!

up to 22 feet
long

The **MOORISH IDOL**'s
a pretty fish.
Of that I have no doubt,
Though it looks a
little strange with that
rather longish snout.
Its snout is used to
reach down into
crevices for its food.
It usually swims in
tiny schools,
which I guess is
pretty shrewd.

N

The **NAUTILUS**
has gas-filled tubes
to help it rise
and sink.
It lives deep in
water that is just
as black as ink.
It has blue blood,
lays eggs and has
a parrot-looking
beak.

The way it moves around is quite ingenious and unique.
It moves by jet propulsion making it go really fast.
That looks like so much fun to do! I'm sure he has a blast.

In a <u>fishy</u> <u>beauty</u> <u>pageant</u>, **NORTHERN
SEA ROB<u>INS</u>** won't <u>win</u>.
They <u>do</u> not <u>look</u> like <u>robins</u> with
large <u>head</u> and <u>fanlike fins</u>.
When <u>they</u> get <u>scared</u> they're <u>not</u> like <u>us</u>
'cause <u>for</u> help <u>we</u> can <u>shout</u>.
They <u>burrow</u> <u>into</u> <u>sand</u>, with <u>just</u> their
<u>head</u> and <u>two</u> eyes <u>sticki</u>ng <u>out</u>.

The **NARWHAL** <u>has</u> a <u>spiral</u> <u>tusk</u>…to
<u>nine</u> feet <u>long</u> it grows!
None <u>know</u> for <u>sure</u> what it's <u>used</u> for.
But I <u>hope</u>, at <u>least</u>, he <u>knows</u>.

O

The **OCTOPUS** has
eight- long arms that are
lined with suction cups.
It can squirt out
jets of water that make it
shoot straight up.
It has a pouch-like body.
With its large brain it's
quite bright.
It can change skin color
in just an instant;
pale to dark….or
speckled to striped.

up to 40 inches long

To confuse its predators it squirts out all this ink.
That way he can get away
before the thing has time to blink.

The **OYSTER
CATCHER**
opens shells
with its bright
red beak.
Its babies stays
home one full
year to learn the
strange
technique.

One species is on the endangered list, another is extinct.
Others, I'm afraid, are close…balancing on the brink.

up to 3 feet long

Sea **OTTERS** spend most of their lives
out there on the sea,
Often swimming among huge beds
of kelp or through seaweed.
They often use rocks like a hammer so
seashells they can crack.
They eat them there, right on the water,
while lying on their backs.
At night they wrap kelp fronds around themselves
because it keeps
Them from drifting out to sea while they are fast asleep.

The **OARFISH** is
unusual
and I am not kiddin'.
Its body's very long
and thin and looks
much like a ribbon.
The way it swims is
like a serpent
rippling back
and forth.
It's found in

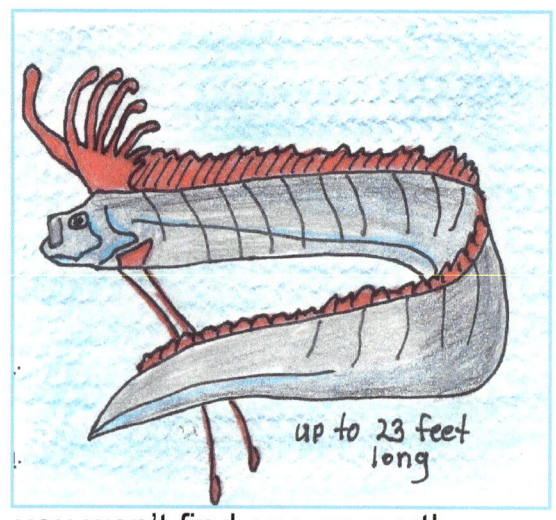

up to 23 feet
long

tropical waters so you won't find one up north.

P

The **PANTHER GROUPER'S** spots are like an underwater camouflage. The flickering sunlight filters down and makes a lovely shadow

collage.
And so the Grouper's spots blend in with the sunspot all around
And it is somewhat hidden. It is not easily found.

Startling a
PUFFER FISH
I would not think
is wise.
It puffs up by
gulping water,
swelling more
than twice
its size.
It looks like a
butterball or
like a prickly
golf ball.

When startled, spines that lie quite flat
then stand up very straight and tall.

Seawater <u>may</u> look
<u>clear</u> to <u>us</u> but
<u>it's</u> filled <u>up</u> with
PLANKTON.
Plankton's <u>made</u> of
<u>tiny</u> <u>life</u>forms <u>and</u> I'll
<u>try</u> to <u>rank</u> them.
<u>Some</u> are <u>plants</u> that
<u>are</u> quite <u>small</u>…
too <u>small</u> for <u>us</u> to <u>see</u>.
To <u>grow</u> and <u>make</u>
new <u>plant</u> material,
they <u>use</u> sun's <u>energy</u>.
The <u>rest</u> are <u>tiny</u> <u>creatures</u>
<u>that</u> on <u>those</u> plants
<u>tend</u> to <u>feed</u>.
Huge <u>things</u> like <u>whales</u>
feed <u>on</u> them <u>both</u>.
That's <u>odd</u>! You <u>must</u> <u>agree</u>!

The **PADDLE WORM**
lives <u>under</u> <u>rocks</u>
<u>and</u> <u>among</u> seaweed.
It <u>uses</u> <u>its</u> four <u>tentacles</u>
to <u>catch</u> its <u>fellow</u>
<u>worms</u> to <u>eat</u>.

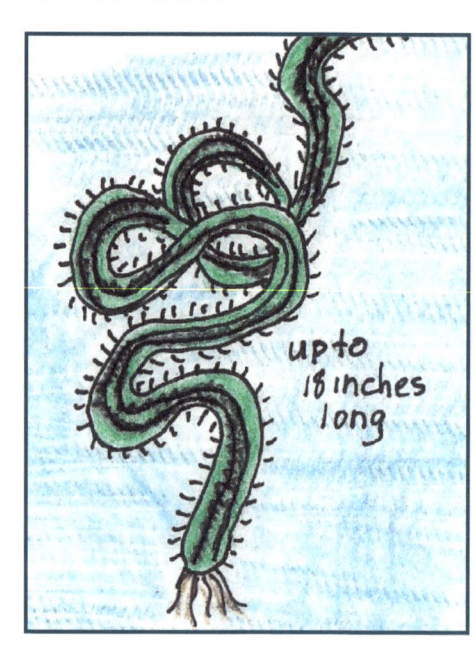

up to
18 inches
long

Stay away from the
PORTUGUESE
MAN-OF-WAR, I say
For they have
stinging cells!
They're used to
paralyze their prey.
If they think you are
prey and sting you,
you will surely weep.
Their tentacles are
very long,
up to sixty feet.
It is a colony of a
hundred individual
critters,
Each one with a
job to do, that are all

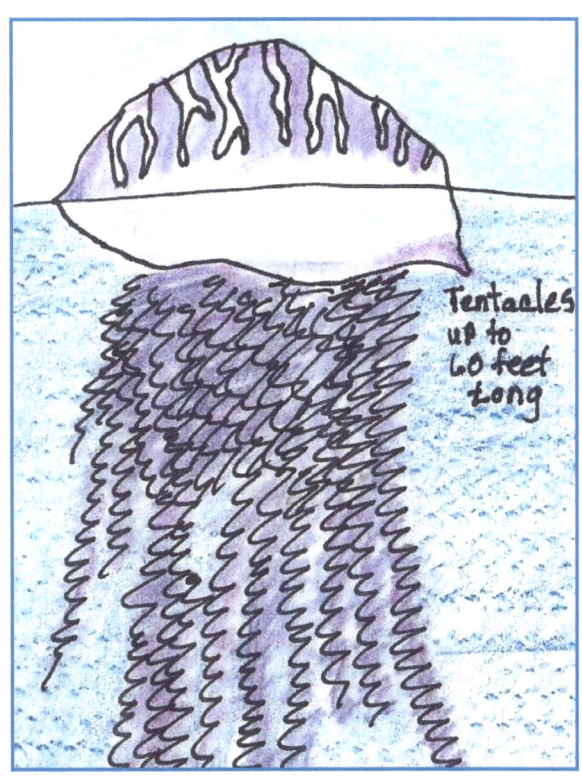

Tentacles up to 60 feet long

joined together.

Q

18 inches long

**QUEEN
ANGELFISH**
provide a
cleaning service
if they wish.
They pick and eat
the parasites that
bother other fish.
Though they are
brightly colored fish, they are quite hard to see
Because they hide among the many colors on a reef.

40.

QUEEN <u>TRIGGERFISH</u>
can <u>be</u> blue, <u>purple</u>,
turquoise, <u>green</u>…
I <u>am</u> impressed!
Their <u>color</u> <u>changes</u> to
<u>what's</u> <u>around</u> them
<u>if</u> they're <u>under</u> <u>stress</u>.
A<u>mong</u> the <u>many</u> <u>colors</u> <u>on</u>
a <u>coral</u> <u>reef's</u>
where <u>they</u> be<u>long</u>.
<u>They</u> can <u>grow</u> to <u>two</u> feet
But they're <u>mostly</u> <u>one</u> foot <u>long</u>.

22 inches long

R

5 feet long

The **<u>RATFISH</u>** <u>gets</u> its
<u>name</u> be<u>cause</u> its
<u>tail</u> looks <u>like</u> a <u>rat's</u>.
It <u>can</u> re<u>lease</u> a
<u>poison</u> <u>so</u> it <u>does</u> not
<u>take</u> much <u>flack</u>.
It <u>swims</u> close <u>to</u>
the <u>seabed</u>, <u>on</u> its
swimming <u>skills</u> <u>relying</u>.
It <u>can</u> do <u>barrel</u> <u>rolls</u> and <u>corkscrew</u> <u>turns</u>
like <u>if</u> it's <u>flying</u>.

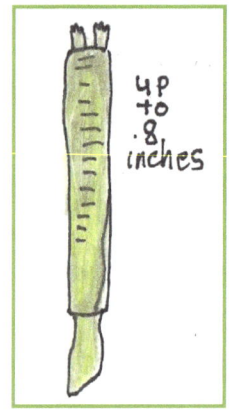

up to .8 inches

The **<u>RAZOR</u> <u>CLAM</u>** looks
<u>kind</u> of <u>like</u> a
razor <u>when</u> you <u>shave</u>.
Its <u>narrow</u> <u>shell</u> can
burrow <u>quickly</u>,
<u>if</u> its <u>life</u> it <u>wants</u> to <u>save</u>.

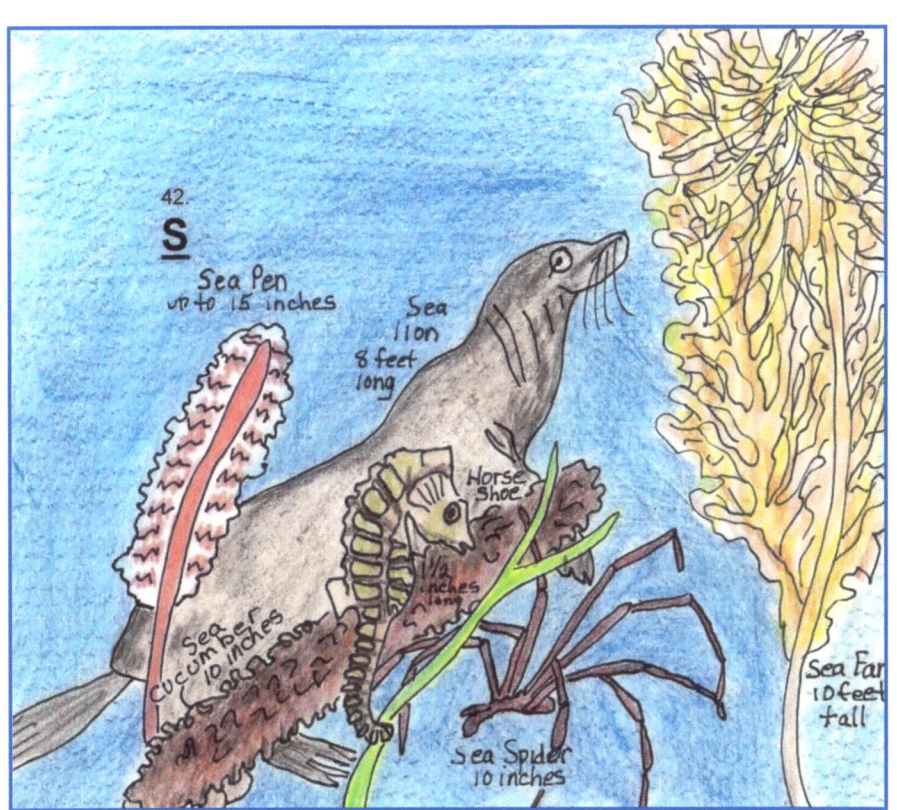

There are <u>many</u> <u>different</u> <u>creatures</u> <u>starting</u> <u>out</u> with '**SEA**'
Like **SEA** PIN, **SEA** SQUIRT, **SEAHORSE**, **SEA** MOUSE
and of <u>course</u> **SEAWEED**.
There <u>are</u> **SEA** **SPIDERS**, **SEA** **LIONS**,
SEA **LILIES**, and **SEA** **PENS**,
Plus **SEA** **OTTERS**, **SEA** SNAILS,
SEA SLUGS <u>and</u>, of <u>course</u>, **SEA** **FANS**.
<u>Some</u> are <u>pretty</u> <u>small</u> and <u>some</u> are <u>very</u> <u>big</u> in<u>deed!</u>
But <u>what</u> they <u>have</u> in <u>common</u> <u>is</u> that <u>they</u> live <u>in</u> the <u>sea</u>.

The **SAND DOLLAR** is
round and thin; and
looks just like a disk.
Kids like to collect them
'cause they're
so hard to resist.
There is a pretty
flower pattern
on top of each shell.
A critter lives inside.
It's like its own
private hotel.

SCALLOPS have
soft bodies.
They're protected
by two shells,
With a row of
eyes between them.
If you look
you just might tell.
To move, it flaps its
shells together
forcing water out.

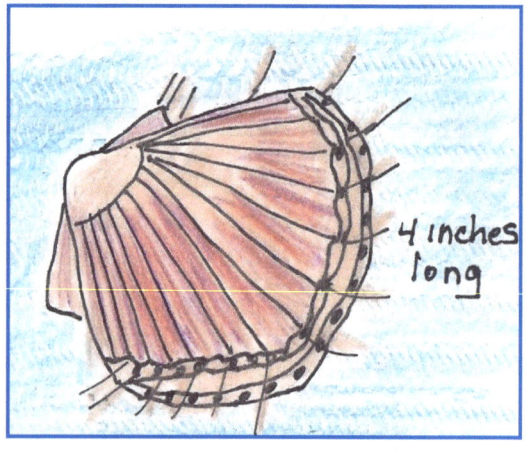

4 inches
long

When it want to go somewhere, that's how it gets about.

The **SAWFISH** has a
nose that
looks just
like a
chainsaw
blade.
It's also
called a
Carpenter
Shark.
Its color is
mostly
gray.

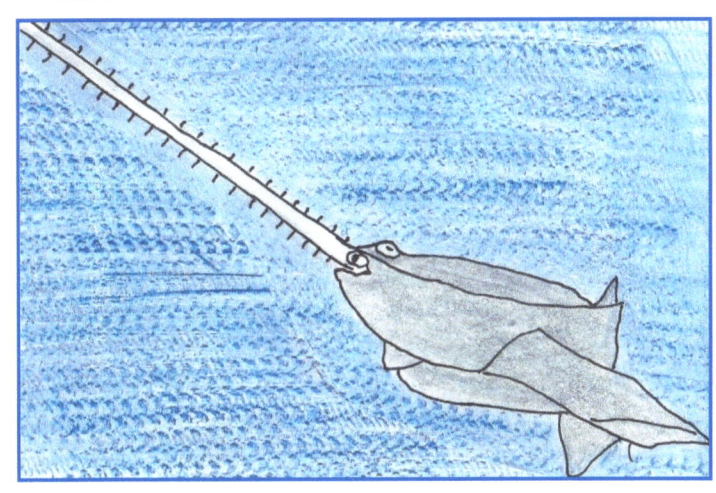

It's critically
endangered, extinction soon to face.
Only five to ten percent are left…what a disgrace!

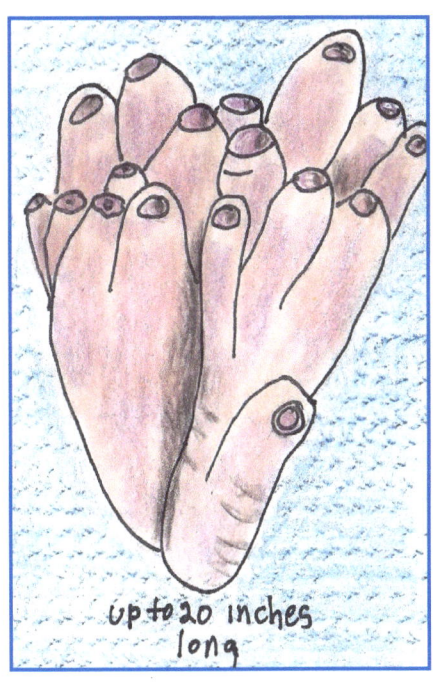

up to 20 inches long

All over its whole body
the **SPONGE** is
filled with holes.
Water flows out through them;
to catch food is its goal.
Sponges come in
many colors,
sizes and weird shapes;
Some as small as thumbnails,
and some as big as apes.
That sponges are among
the simplest animals is plain
Yet they've lived five
hundred million years….
without a brain!

44.

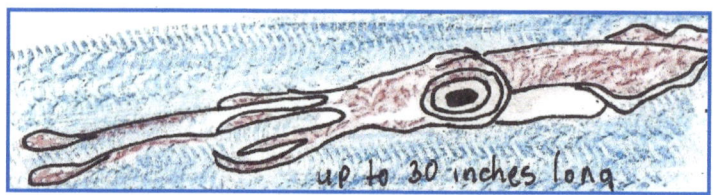

The S**QUID'S** body kind of looks a lot like a torpedo.
It moves by propulsion so you
can't say that it moves slow.
It moves backwards very quickly
like a torpedo does.
Here it comes! Here it is! Here is where it was!

The **SHRIMP'S** outer
skeleton
is lighter than a crab's.
It has so many
limbs that it'd be
difficult to add!
Some are used for
swimming;

Some to help it eat;
And some are used for walking on the bottom of the sea.

Baby **SWORDFISH**
do not have a
long snout sticking out.
The sword grows
as the fish grows, but
slowly, not flat-out.
Sixteen feet's as
big as these fish

typically will grow
So I am really glad that their sword snouts
tend to grow slow.

SALMON live in the
deep ocean from two
years up to four.
Then they head for the
breeding ground, the
very spot where they
hatched before.
Sometimes they'll
travel a
thousand miles to
reach that very spot.
How they know just
where to go
is knowledge
I've not got.

Because **SQUIRRELFISH** live in the dark
their eyes are large, I think.
Their fins and tails are silver gray but their
body's a reddish pink.
All day they hide under rocky ledges,
in crevices or in caves.
It's because they are nocturnal.
That's how their kind behaves.

T

TURTLE flippers beat like wings as gracefully it glides.
Most of its life is spent in water and very little outside.
The female lays her eggs ashore into sand- buried piles.
At times, to lay her eggs, she'll travel
many hundred miles.

The **TUNA FISH** is very fast
because it's shaped for speed.
It can swim to fifty miles per hour, yes indeed!
It hardly ever stops because to move's a basic need.
For while he swims his mouth is open.
The water helps him breathe.
The biggest was ten-feet and weighed
one thousand pounds or more!
Some are called the Bluefin, Bigeye,
Skipjack and Albacore.

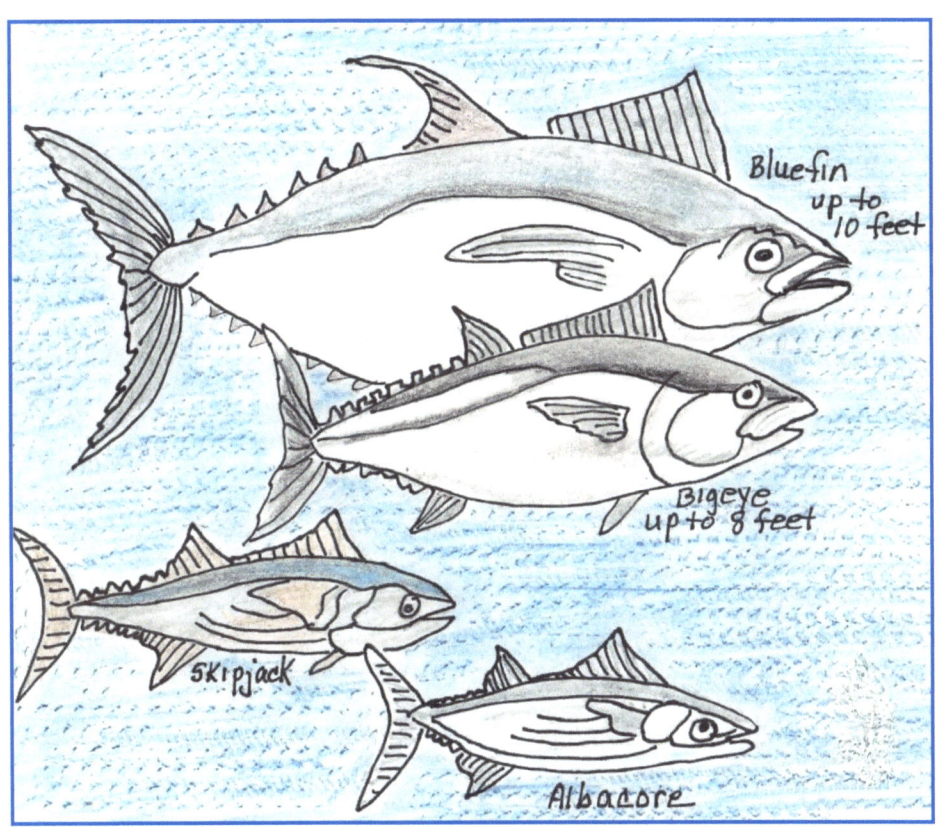

4 inches wide.
Spines 4 to 16 inches long

U

The Long-spine **URCHIN**
has a body
that's four inches wide.
But the spines, some
sixteen inches long,
are like pins
stuck in its side.
Its circle of teeth is
used for chewing but it
just does that at night,
For all day long it
tries to hide,
completely out of sight.

V

The **VIPERFISH** can
open its jaws
extremely wide, indeed.
Its teeth are long
and needle-like.
It punctures its prey,
then eats.
It swallows them whole!
It never chews its
prey or takes a bite.
It lives deep down
where it is dark so
carries a glowing light.

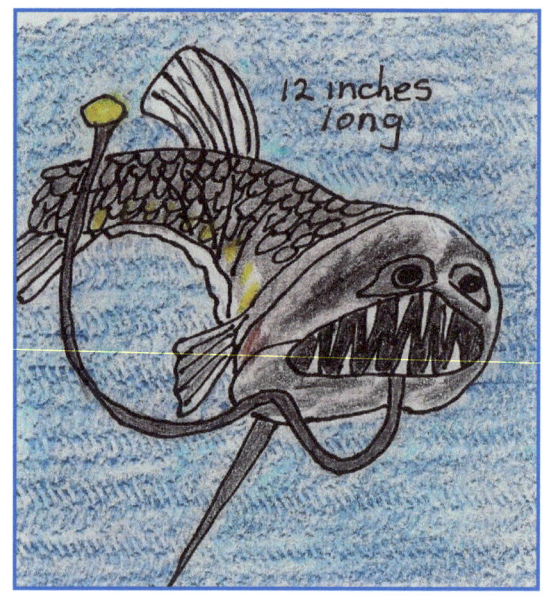

12 inches long

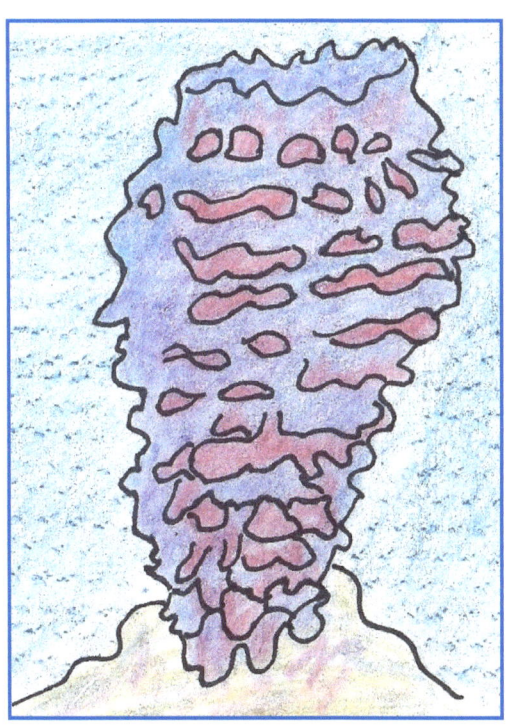

The **VASE SPONGE**
draws up water for
its wall has tiny holes.
To gather up organic
plankton is its
one true goal.
It really does look
like a pretty
decorated vase.
I'm tempted to put
flowers in it
and give it away!

W

In shallow coastal water
where the sunlight
can be felt,
Down in the sandy
sea bottom
lives the common
WHELK.
The water there is clear
and so the sunlight
rarely fades.
The seaweed is

16 inches

quite plentiful and is chopped up by the waves.
Algae is abundant too because of the sun's heat
And so the Whelk is happy for there's plenty there to eat.

50.

The **WRASSE**
is found
all over the world
in water that
is warm.
Adult males have

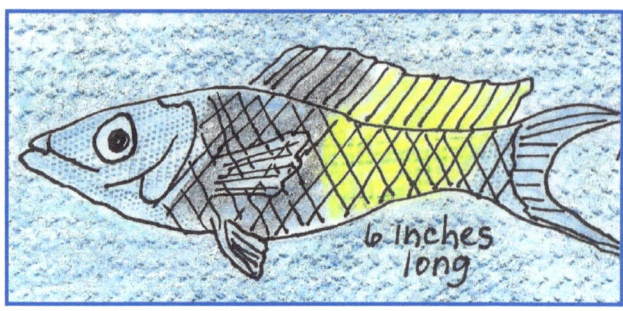

bright colors but the rest do not conform.
Females and young males are mostly yellowish, not drab.
They use their teeth to crush their prey,
their favorite being crab.

The largest fish in the whole world
is the huge **WHALE SHARK**
He's quite striking on the top but
quite plain on his underparts.
He can sometimes grow quite long – as much as 60 feet.
That's like a three-story building!
I doubt that can be beat.
A fierce hunter he is not (for that humongous size).
He leaves large fish alone and goes
after the small-fries.

Different **WHALES** are different lengths,
from sixteen to one hundred feet.
Most go after tiny Krill.
By the tons these they will eat.
In stories sometimes characters are
swallowed by huge whales.
But unless they were krill-size
that seems to me a fairy-tale.

 up to 11½ feet long

The **WALRUS** is
huge and heavy
but
in water he is quick.
His fatty blubber
keeps him warm for
it's four inches thick!
His tusks, that grow
to
three feet long,
are used,
its understood,

To dig up mollusks on the seafloor
when he looks for food.

Y

The **YELLOW-
TAIL
SNAPPER**
lives
close to the
coral reefs.
It is a tasty fish
so it is

24 inches long

one that humans eat.
It is a pretty fish with its bright yellow tail and stripe,
Though most of the rest of it is just a plain, old white.

Z

You'll find
ZOANTHARIA
in deep seas
or coral reefs.
It is highly poisonous
but just…
if it you eat.
It uses
photosynthesis,
like plants, to make
its food.
But a diet with some
plankton they
too must include.

So here you have some of the critters
found around the sea.
I could not include them all
because there's just too many.
So if you want to learn more, then
some library books check out.
There you'll learn a whole lot more.
Of this I have no doubt.

LOOK FOR OTHER BOOKS
FROM THE
A-B-C SCIENCE SERIES:

The A-B-C Seashell Book
The A-B-C Tree Book
The A-B-C Dinosaur Book
The A-B-C Wildflower Book
The A-B-C Bird Book
The B-B-C Bug Book
The A-B-C Butterfly Book
The A-B-C Zoo Book
The A-B-C Body Book
The A-B-C Amphibians and Reptile Book
The A-B-C Big Cats, Little Cats Book
The A-B-C Tropical Fish Book
THE A-B-C- DOG BOOK

www.ingramcontent.com/pod-product-compliance
Lightning Source LLC
Chambersburg PA
CBHW040856180526
45159CB00001B/441